WHO EATS WHAT?
RAIN FOREST
FOOD CHAINS

by Rebecca Pettiford

pogo

Ideas for Parents and Teachers

Pogo Books let children practice reading informational text while introducing them to nonfiction features such as headings, labels, sidebars, maps, and diagrams, as well as a table of contents, glossary, and index.

Carefully leveled text with a strong photo match offers early fluent readers the support they need to succeed.

Before Reading

- "Walk" through the book and point out the various nonfiction features. Ask the student what purpose each feature serves.

- Look at the glossary together. Read and discuss the words.

Read the Book

- Have the child read the book independently.

- Invite him or her to list questions that arise from reading.

After Reading

- Discuss the child's questions. Talk about how he or she might find answers to those questions.

- Prompt the child to think more. Ask: What other rain forest animals and plants do you know about? What food chains do you think they are a part of?

Pogo Books are published by Jump!
5357 Penn Avenue South
Minneapolis, MN 55419
www.jumplibrary.com

Library of Congress Cataloging-in-Publication Data

Pettiford, Rebecca, author.
  Rain forest food chains: who eats what? / by Rebecca Pettiford.
    pages cm – (Who eats what?)
  Audience: Ages 7-10.
  Includes index.
  ISBN 978-1-62031-304-6 (hardcover: alk. paper) –
  ISBN 978-1-62031-431-9 (paperback) –
  ISBN 978-1-62496-356-8 (ebook)
  1. Rain forest ecology–Juvenile literature. 2. Food chains (Ecology)–Juvenile literature. 3. Rain forest animals–Juvenile literature. I. Title.
  QH541.5.R27P48 2016
  577.34–dc23
                                        2015028824

Series Editor: Jenny Fretland VanVoorst
Series Designer: Anna Peterson
Photo Researcher: Anna Peterson

Photo Credits: All photos by Shutterstock except: ardea, 20-21bm; Corbis, cover; iStock, 9; Nature Picture Library, 16-17, 19; SuperStock, 12-13, 14-15; Thinkstock, 20-21b.

Printed in the United States of America at Corporate Graphics in North Mankato, Minnesota.

TABLE OF CONTENTS

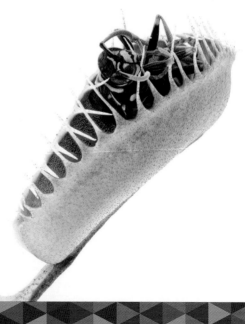

CHAPTER 1

A WALK IN THE RAIN FOREST

Let's walk in a **tropical** rain forest. This **biome** gets more rain than any place on Earth.

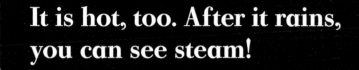

It is hot, too. After it rains, you can see steam!

The forest is thick with tall trees. The forest floor does not get a lot of sun. Listen. Can you hear the animals? Rain forests are home to more than half of the animals that live on Earth.

WHERE ARE THEY?

Tropical rain forests are in South America, Africa, and Asia.

■ = Tropical Rain Forests

CHAPTER 2

THE RAIN FOREST FOOD CHAIN

Plants and animals need energy to grow and move. Food is energy. Plants use the sun, soil, and water to make food. Animals eat plants and each other.

A **food chain** describes the way energy moves through a series of living things. Each food chain begins with plants and ends with animals. Each link in the food chain uses or eats the one that comes before it.

Tree leaves, fruit, and grass are **producers**. They are the first link in the food chain. These plants use energy from the sun to make their own food. The rain and heat make it possible for many kinds of plants to grow.

DID YOU KNOW?

Some rain forest plants eat animals. Venus flytraps and pitcher plants eat insects and even frogs!

toucan
(consumer)

tropical fruit
(producer)

Insects, birds, and other animals eat the plants. They are **consumers**, the next link in the chain. There are many consumers in the rain forest. Bats, monkeys, and **sloths** are just a few. People eat rain forest plants, too. We eat nuts, spices, and tropical fruit.

DID YOU KNOW?

Sloths have tiny green plants growing on their fur. This helps them blend in with the forest. Moths live in their fur, too. The moths eat the tiny plants, and the fur helps them hide from birds.

Jaguars and giant snakes are the next link in the food chain. These **predators** eat consumers. They also eat smaller predators.

DID YOU KNOW?

Animals will eat different things to live. This means they are part of more than one food chain. When food chains cross, they make a food web.

tree boa
(predator)

Whether consumer or predator, all animals eventually die. At this point, **decomposers** such as **millipedes** break down the body. The dead matter turns into nutrients, which return to the forest.

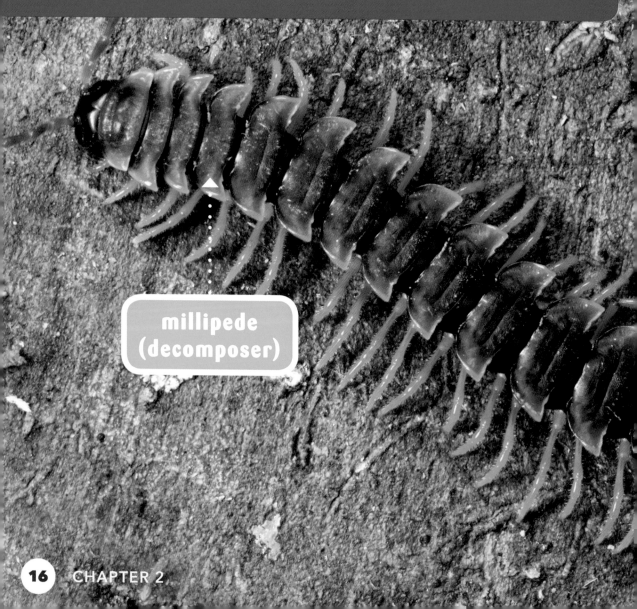

millipede
(decomposer)

· ·

One rain forest food chain might look something like this:

Producer:
Tropical Plant

Predator:
Jaguar

Consumer:
Monkey

Decomposer:
Millipede

FOOD CHAIN CLOSE-UPS

Let's look at a simple food chain. A banana grows on a tree. A monkey eats the banana.

An **ocelot** eats the monkey. A giant snake eats the ocelot. When the snake dies, decomposers break down its body.

Let's look at another food chain.

1) A tree grows leaves.

2) A sloth eats the leaves.

3) A **harpy eagle** eats the sloth.

4) In time, the eagle dies. Insects break down the eagle's body.

Nutrients return to the forest. The food chain continues!

ACTIVITIES & TOOLS

TRY THIS!

YOU CAN HELP!

We need rain forests to keep our planet healthy. Rain forests trees make air so we can breathe. Scientists use the plants to make drugs that fight sickness.

But rain forests and their food chains are in danger. Each year, people clear more trees to make farms and paper and wood products.

You can help save the rain forests.

Here are some ideas:

1. Reuse paper instead of throwing it out.

2. Don't buy things made from rain forest woods. These include mahogany, rosewood, and ebony.

3. Hold a bake sale or yard sale. Give the money to groups that save rain forests.

4. Ask friends and family to honor your birthday with a donation to the Rainforest Foundation: www.rainforestfoundation.org

5. Ask your parents to buy bananas and coffee that are grown in a way that is safe for rain forests.

6. Avoid foods with palm oil. Rain forests are cut down to grow palm trees for the oil.

GLOSSARY

biome: A large area on the earth defined by its weather, land, and the type of plants and animals that live there.

consumers: Animals that eat plants.

decomposers: Life forms that break down dead matter.

food chain: An ordering of plants and animals in which each uses or eats the one before it for energy.

harpy eagle: A big eagle that lives in tropical rain forests.

jaguars: Large cats that live in the rain forests of South America.

millipedes: Worm-like decomposers.

ocelot: A medium-sized cat that has a yellow coat with black spots.

predators: Animals that hunt and eat other animals.

producers: Plants that make their own food from the sun.

sloths: Tropical, slow-moving animals that hang upside down from the branches of trees.

tropical: An area of the world with hot, wet weather.

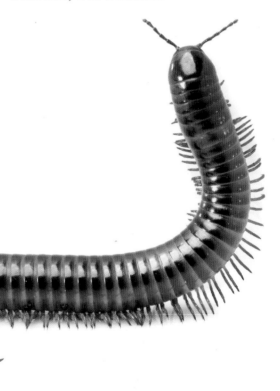

INDEX

TO LEARN MORE

Learning more is as easy as 1, 2, 3.

1) Go to www.factsurfer.com

2) Enter "rainforestfoodchains" into the search box.

3) Click the "Surf" to see a list of websites.

With factsurfer, finding more information is just a click away.